FLORA

Based on *The Railway Series* by the Rev. W. Awdry

Illustrations by
Robin Davies and *Jerry Smith*

EGMONT

EGMONT

We bring stories to life

First published in Great Britain in 2009
by Egmont UK Limited
239 Kensington High Street, London W8 6SA

Thomas the Tank Engine & Friends™

CREATED BY BRITT ALLCROFT

Based on the Railway Series by the Reverend W Awdry
© 2009 Gullane (Thomas) LLC. A HIT Entertainment company.
Thomas the Tank Engine & Friends and Thomas & Friends are trademarks of Gullane (Thomas) Limited.
Thomas the Tank Engine & Friends and Design is Reg. U.S. Pat. & Tm. Off.

HiT entertainment

ISBN 978 1 4052 4422 0
1 3 5 7 9 10 8 6 4 2
Printed in Italy

FSC
Mixed Sources
Product group from well-managed
forests and other controlled sources
Cert no. TT-COC-002332
www.fsc.org
© 1996 Forest Stewardship Council

Egmont is passionate about helping to preserve the world's remaining ancient forests.
We only use paper from legal and sustainable forest sources.

This book is made from paper certified by the Forestry Stewardship Council (FSC),
an organisation dedicated to promoting responsible management of forest resources.
For more information on the FSC, please visit www.fsc.org. To learn more about
Egmont's sustainable paper policy, please visit www.egmont.co.uk/ethical

This is a story about Flora, a new steam tram on the Island of Sodor. Thomas thought that Toby would be upset if he knew there was another tram engine on my railway. But Flora soon proved how Really Useful she could be …

Thomas and Toby are very good friends. One bright morning, Thomas puffed past Toby's shed.

"Hello, Toby!" he peeped. "You look happy."

"The Fat Controller has asked me to lead the parade at the first Great Waterton Festival!" Toby wheeshed, excitedly.

Thomas was pleased for his friend. "Everyone will cheer for you, Toby. You're the only steam tram on Sodor!" he smiled.

Thomas puffed quickly away. The Fat Controller was waiting for him at Great Waterton.

When Thomas arrived in Great Waterton, he stopped suddenly in his tracks.

"Fizzling fireboxes!" he gasped.

There, on the road next to the track, was *another* steam tram.

"Thomas," boomed The Fat Controller. "This is Flora. She is the new steam tram on Sodor."

Flora smiled, happily. "Hello, Thomas!" she puffed.

"H-h-hello," Thomas puffed back.

"Flora is to lead the parade with Toby," The Fat Controller said. "Please take her to meet him. Then, Toby can bring Flora back here, in time for the parade!"

Flora's crew changed her road wheels for steel railway ones so that she could travel on the track.

"Toby thinks he's leading the parade all by himself," he worried, quietly. "I must keep Flora away from Toby until after the parade!"

Thomas knew he would get into trouble with The Fat Controller, but didn't want Toby to be upset.

Before long, Thomas and Flora were steaming through the countryside. When they had to stop at a signal, an idea flew into Thomas' funnel.

"I will take Flora to do one of my jobs before we go to see Toby," he chuffed to himself. "That way, Toby is sure to have set off to lead the parade!"

"Let's go to the Wood Yard first," Thomas puffed to Flora. "It won't take long!"

"That sounds splendid," Flora smiled, sweetly. And she steamed after him.

At the Wood Yard, Thomas collected a flatbed of logs. "Now, we can go and meet Toby!" he chuffed.

Flora was pleased. "Hooray!" she peeped.

"Toby, Toby, you must go. Hurry now to lead the show!" Thomas puffed to himself, as they set off.

Thomas and Flora soon reached the junction by Toby's shed. The shed looked empty.

They were about to go again when Thomas saw steam coming from inside. "Cinders and ashes!" Thomas wheeshed, quietly. "Toby's still there! He mustn't see Flora!"

Thomas puffed off as fast as his wheels would carry him.

"Quickly, Flora!" he whistled, loudly. "I've just remembered. We must go to the Quarry first!"

Flora was surprised! "If you say so, Thomas," she chuffed, chirpily.

And she followed Thomas down a narrow track to the Quarry.

At the Quarry, Thomas coupled up to some slate trucks.

"*Now*, we can go to meet Toby," he told Flora.

Flora was excited.

"*Toby, Toby, please have gone. Lead the show, be proud and strong!*" he puffed to himself.

The two engines went back to Toby's shed. This time, Thomas was sure it was empty.

Suddenly, there was a 'ding, ding' sound. It was Toby's bell!

"Oh, no!" thought Thomas. "Toby's still inside his shed! He mustn't see Flora!"

"Quickly, Flora," Thomas wheeshed. "We're late to pick up my load at the Docks."

Flora was surprised. "If you say so, Thomas!" she said, wearily. And they clickety-clacked back down the track.

At the Docks, Thomas was coupled up to his large load to take to Great Waterton.

Thomas was sure Toby would be leading the parade by now! "Come on, Flora!" he whistled.

Flora's axles ached. Then there was trouble . . .

Flora had chuffed too far . . . and run out of coal!

"I'm sorry, Thomas," Flora cried. Her wheels wobbled weakly, as she came to a stop.

Thomas felt sorry. He knew it was his fault. But just then, he heard a chuff . . . and a puff . . . and the 'ding, ding' of a bell.

It was Toby! "Thomas, I've been waiting for you!" Toby huffed. Then he saw Flora. "Who's that?" he gasped.

Thomas was worried. Neither of them was at the parade. The Fat Controller would be very cross!

"Flora, this is Toby. Toby, this is Flora!" Thomas puffed, bravely.

But instead of being upset, Toby smiled. "Oh, Thomas! I wanted to tell you that I was too scared to lead the parade all by myself. Now, Flora and I can lead it together!"

Thomas had to put right his mistake, and quickly.

"Flora, take some of my coal," Thomas chuffed. "I'll collect more from the coal shed!"

So Thomas' crew took some of his coal and made a good fire in Flora's firebox.

Before long, Flora felt much better.

"Now puff as fast as you can to Great Waterton!" Thomas told the two tram engines.

Toby and Flora chuffed quickly away.

Thomas puffed into Great Waterton just in time to see the flatbed being unloaded. It was a wonderful tram car for Flora!

Flora gasped. Toby rang his bell, 'ding, ding'.

And Thomas smiled, as his two friends set off, leading the first ever Great Waterton Parade.

The children cheered and the brass band boomed. Toby and Flora looked splendid!

"Now I have two best friends," Thomas peeped. "And they are *both* steam trams!"

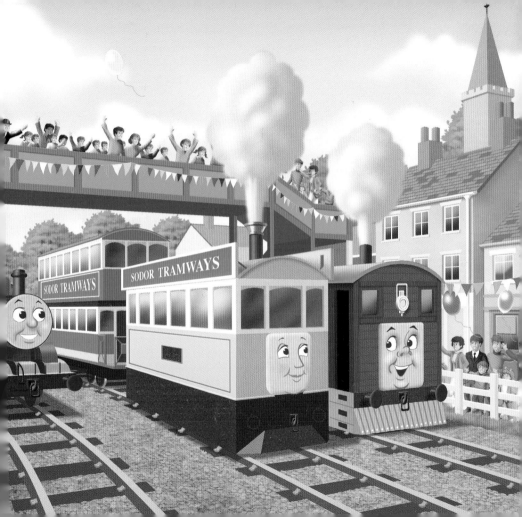

Two Great Offers for Thomas Fans!

THOMAS & FRIENDS

In every Thomas Story Library book like this one, you will find a special token. Collect the tokens and claim exclusive Thomas goodies:

Offer 1

Collect 6 tokens and we'll send you a **poster** and a **bookmark** for only **£1**.
(to cover P&P)

Reply Card for Thomas Goodies!

1 Yes, please send me a **Thomas poster and bookmark.** I have enclosed **6 tokens plus a £1 coin** to cover P&P. ☐

2 Yes, please send me a **Thomas book bag.** I have enclosed **12 tokens plus £2** to cover P&P. ☐

Simply fill in your details below and send them to:
Thomas Offers, PO BOX 715, Horsham, RH12 5WG

Fan's Name: ....

Address: ....

....

.... Date of Birth: ....

Email: ....

Name of parent/guardian: ....

Signature of parent/guardian: ....

Please allow 28 days for delivery. Offer is only available while stocks last. We reserve the right to change the terms of this offer at any time and we offer a 14 day money back guarantee. This does not affect your statutory rights. Offer applies to UK only. The cost applies to Postage and Packaging (P&P).

We may occasionally wish to send you information about other Egmont children's books but if you would rather we didn't please tick here ☐